The Activity and Use of Dietary Zinc in Poultry Nutrition and Pharmacological Roles

The Activity and Use of Dietary Zinc in Poultry Nutrition and Pharmacological Roles

Reza Vakili

ELIVA PRESS

ISBN: 978-99949-8-225-7

Cover Design: Eliva Press
Cover Image: Ingimage
Email: info@elivapress.com
Website: www.elivapress.com

This Eliva Press imprint is published by the registered company Eliva Press Global Ltd. part of Eliva Press S.R.L. Publishing Group

The registered company address is: Pope Hennessy Street Level 2, Hennessy Tower Port Louis, Mauritius
Eliva Press S.R.L. Publishing Group address is: Bulevardul Moscova 21, Chisinau, Moldova, Europe

The activity and use of dietary zinc in poultry nutrition and pharmacological roles

ABSTRACT

Zinc (Zn) is an important trace mineral which has various roles in poultry nutrition. It serves as an essential nutrient and a metal cofactor for several enzymes. Sufficient zinc intake and absorption are required for various functions such as immunity against pathogens, growth, and quality of the egg. It is now recognised that Zn has a major role in antioxidation, growth and development, production, immunity and stress related issues. Supplementation of Zn can improve growth, augment immunity, enhance antioxidant capacity, increase endocrine secretion and interact with other minerals in the gut.While zinc's importance as an essential nutrient has been recognized for many years, only recently have researchers understood the full impact of this nutrient on animal and human health. Researchers have identified over 200 zinc-dependent enzymes in all the major biochemical pathways in the body. Zinc is an essential component of both DNA and RNA polymerase enzymes. It is vital to the activity of a variety of hormones including glucagon, insulin, growth hormone, and the sex hormones. It also plays a key role in the immune system.

There have been some reports indicating that supplementation of Zn could alleviate the negative effects of age on egg quality in laying hens.However,information regarding these positive effects on health,bone mechanical properties and zinc deposition in body is limited. In this book, some of the basic information regarding the importance and activity of Zn within the body and the addition of Zn in poultry diets is discussed. This article provides an update for nutrition professionals on the current understanding of both the physiological and pharmacological roles of zinc in poultry diets.

Table of Contents

Chapter 1: Our current understanding of both the physiological and pharmacological roles of zinc in animal diets

Introduction :

Appetite:

Reduced appetite is one of the first zinc deficiency signs observed in animals. In rats, zinc deficiency often reduces feed intake by 30%, and force-feeding the deficient rats rapidly induces illness. O'Dell and Reeves (1989) showed that changes in appetite are associated with changes in the concentrations of amino acid derived neurotransmitters in the brain. Zinc-deficient rats change their dietary preferences, avoiding carbohydrates and seeking protein and fat (Kennedy et al., 1998). Key enzymes required for carbohydrate metabolism may be lacking because the zinc dependent messenger RNA needed to synthesize these enzymes has reduced expression. The rate at which zinc- deficient rats respond to zinc supplementation is amazing. Chester and Quarterman (1970) showed that the rate of food intake increased within 1-2 hours of zinc supplementation.

Zinc deficiency may also reduce appetite by impairing of taste. The sense of taste is mediated through the salivary zinc dependent polypeptide, gustin. Low salivary zinc concentration leads to a reduction of taste and reduced appetite. Droke et al. (1993) showed that as lambs became zinc deficient their eating behavior changed. As their zinc status declined from adequate to deficient they regressed from meal-eaters to nibblers. It is hypothesized that the reduction in enzyme activity leads to the accumulation of one or more metabolites causing a marked change in eating behavior. This extreme sensitivity of appetite to nutrient supply is unique to zinc, expressed in all species, and reflects the key role of zinc in nutrient metabolism.

Immunocompetence:

Zinc deficiency causes decreased immunity and loss of T-cell function in animals. Splenic marcrophages from zinc deficient mice were less able to facilitate T-cell mitogensis than from pair-fed controls and was directly related to the degree of zinc depletion (James et al., 1987). Many researchers have assumed that the decreased immune response is a secondary response associated with reduced nutrient intake. Droke (1993) proposed that loss of appetite, poor growth, and skin lesions occurred before increased susceptibility to infection in zinc deficient lambs. In contrast, Engle et al. (1997) reported that the responses to a subcutaneous injection of phytohaemagglutinin were impaired by zinc deprivation before there was any loss of appetite or drop in plasma zinc in heifers.

Skin and Skeletal Signs:

Parakeratosis, the thickening, hardening and cracking of skin is a common sign of zinc deprivation in all species. Poultry typically develop a severe dermatitis of the feet and poor feathering. Pigs develop skin lesions over the extremities. Calves show similar symptoms over the neck, ears and hind limbs. Zinc deficiency greatly retards the rate of healing of skin wounds in all species. In sheep, wool looses its crimp and the whole fleece may be shed with a severe deficiency. Zinc is critical to proper skeletal growth during embryonic development. Calves born to zinc deficient dams have exhibited bowing of the hind limbs, stiffness of joints, and swelling of the hocks (Miller and Miller 1962). Zinc is critical to good hoof health in many species. Zinc improves hoof health through keratin synthesis and maturation, wound healing, and epithelium maintenance.

Reproduction:

Decreased litter size was one of the first observations with zinc deficient pigs (Hoekstra et al., 1967). In poultry, decreased egg hatchability is associated with zinc deficiency. In ewes, even mild zinc deficiency has reduced both birth weight and the number of offspring (Masters and Fels, 1980). In rams spermatogenesis dropped to almost zero after 20 weeks on a diet containing 2.4 ppm of zinc. Rams typically require approximately 30 ppm zinc for optimal fertility (Underwood and Somers, 1969). White (1993) reported that zinc deficiency-induced anorexia reduced the secretion of gonadotrophin-releasing hormone, a key reproductive regulator, in rams. Pregnancy toxemia has also occurred as a secondary consequence of anorexia in the ewe (Apgar et al., 1993).

Pharmacological Role:

Recent data suggest that high levels of zinc in certain diets may improve animal health independent of its role on the immune system. Research conducted by Hahn and Baker (1993), Carlson et al. (1999) and Hill et al. (2000) showed that feeding 3,000 ppm zinc, added as zinc oxide, enhances growth and health of nursery pigs. The nursery period is typically a 28-day period beginning at weaning (15-20 days of age). In general, growth rates have been improved 10-25% with a 0-15% increase in feed intake.

Although it is becoming routine to add 2,000 to 3,000 ppm of zinc to nursery diets, the exact mechanism behind the enhanced performance is unknown. Carlson et al. (1998) reported that feeding 3,000 ppm zinc as zinc oxide produced deeper crypts and greater total thickness in the duodenum. Katouli et al. (1999) found that 2,500 ppm zinc in diets of weanling pigs helped maintain the stability of intestinal microflora and diversity of the coliforms for the first 2-weeks after weaning. Some researchers have speculated that pharmacological zinc levels may improve performance through a systemic effect via the blood rather than the enteric effects in the small intestine. However, recent data suggest that bioavailability of the zinc is irrelevant to the performance response (Case and Carlson, 2002). The plasma, tissue, urine, and fecal zinc concentrations of pigs fed 500 ppm of zinc from two organic sources were not different compared to zinc oxide. These data would lead one to conclude that there were no differences in bioavailability between the organic zinc and zinc oxide sources. Additional research is needed to fully understand the mode of action.

Conclusion :

Although the importance zinc as a nutrient has been known for over a century, we are still learning its roles both nutritionally and pharmacologically. Zinc is critical to maintaining appetite, reproductive efficiency, skin and skeletal health, a strong immune system, and a host of other functions as a cofactor for over 200 enzymes. More recently the pharmacological role of zinc as a feed additive for nursery pigs has been demonstrated. Using a trace mineralized salt that is well fortified with bioavailable zinc is the foundation for maintaining the performance, health and vitality of livestock and poultry.

Literature Cited :

Apgar, J., G.A. Everett and J.A. Fritzgerald. 1993. Dietary zinc deprivation effects parturition and outcome of pregnancy in the ewe. Nutrition Research 13:319.

Carlson, M.S., G.M. Hill, and J.E. Link. 1999. Early and traditionally weaned nursery pigs benefit from phase-feeding pharmacological concentrations of zinc oxide: Effect on matallothionein and mineral concentrations.

Carlson, M.S., S.L. Hoover, G. M. Hill, J.E. Link, and J.R. Turk. 1998. Effect of pharmacological zinc on intestinal metallothionein concentration and morphology of the nursery pig. J. Anim. Sci. 76(Suppl.1):57 (Abstr.).

Case, C.L., and M.S. Carlson. 2002. Effect of feeding organic and inorganic sources of additional zinc on growth performance and zinc balance in nursery pigs. J. Anim. Sci. 80:1917.

Chester, J.K., and J. Quarterman. 1970. Effects of zinc deficiency on food intake and feeding patterns of rats. Brit. J. of Nutrition. 24:1061.

Droke, E.A., J.W. spears, T.T. Brown, and M.A. Quereshi. 1993. Influence of dietary zinc and dexamethason on immune responses and resistance to Pasteurella haemolytica challenge in growing lambs. Nutrition Research 13:1213.

Droke, E.A., J.W. Spears, J.D. Armstrong, E.B. Kegley, and R. Simpson. 1993. Dietary zinc affects serum concentration of insulin and insulin-like growth factor I in lambs. J. of Nutrition. 123:13.

Engle, T.E., D.F. Nockels, C.V. Kimberling, D.L. Weaber, and A.B. Johnson. 1997. Zinc repletion with organic and inorganic forms of zinc and protein turnover in marginally zinc-deficient calves. J. of Anim. Sci. 75:3074

Hahn, J.D. and D.H. Baker. 1993. Growth and plasma zinc responses of young pigs fed pharmacologic levels of zinc. J. Anim. Sci. 71:3020.

Hill, G.M., G.L. Cromwell, T.D. Crenshaw, C.R. Dove, R.C. Ewan, D.A. Knabe, A.J. Lewis, G.W. Libal, D.C. Mahan, G.C. Shurson, L.L. Southern, and T.L. Veum. 2000. Growth promotion effects and plasma changes from feeding high dietary concentrations of zinc and copper to weanling pigs (regional study). J. Anim. Sci. 78:1010.

Hoekstra, W.G., E.C. Faltin, C.W. Lin, H.F. Roberts, and R.H. Grummer. 1967. Zinc deficiency in reproducing gilts fed a diet high in calcium and its effect on tissue zinc and blood serum alkaline phosphatase. J. Anim. Sci. 26:1348.

James, S.J., M. Swenseid, and T. Makinodan. 1987. Macrophage-mediated depression of T-cell proliferation in zinc-deficient mice. J. of Nutrition. 117:1982.

Katouli, M., L. Meliin, M. Jensen-Waern, P. Walgren, and R. Mollby. 1999. The effect of zinc oxide supplementation on the stability of the intestinal flora with special reference to composition of coliforms in weaned pigs. J. Appl. Microbiol. 87:564.

Kennedy, K.J., T.M. Rains, and N.F. Shay. 1998. Zinc deficiency changes preferred maconutrient intake in subpopulations of Sprague-Dewley outbred rats and reduces hepatic pyruvate kinase gene expression. J. of Nutrition 128:43.

Masters, D.G., and H.E. Fels. 1980. Effect of zinc supplementation on reproductive performance of grazing Merion ewes. Biological Trace Element Research. 7:89.

Miller, J.K., and W.J. Miller. 1962. Experimental zinc deficiency and recovery of calves. J. of Nutrition. 76:467.

O'Dell. B.L., and P.G. Reeves 1989. Zinc status and food intake. In: Zinc in Human Biology. ILSI Press, Washington, DC pp. 173.

Underwood, E.J. and Somers, M. (1969). Studies of zinc nutrition in sheep. 1. The relation of zinc to growth, testicular development and spermatogenesis in young

rams, Australian Journal of Agricultural Research 20:889
White, C.L., 1993. The zinc requirements of grazing ruminants. In: Robson, A.D. (ed.) Zinc in Soils and Plants: Developments in Plant ans Soil Sciences, Vol. 55, Kluwer Academic Publishers, London, pp. 197.

Chapter 2: Recent advances in zinc supplementation in poultry diets

Zinc supplementation improves antioxidant and status bone Health of aged laying hens

Introduction

Zinc is vital for growth and development, egg quality, bone health, and immune function in laying hens (Yu et al.2020) and also antioxidant protection system (Vakili &Rashidi, 2011).Cell growth, differentiation, reproduction, carbohydrate and protein metabolism involve Zn, which has a important role in the isthmus during the formation of the eggshell membranes and in the magnum during the albumen deposition. Zn is considered a cofactor for carbonic anhydrase supplying carbonic ions during shell formation. Carbonic anhydrase inhibitors cause an enhancement in the production of shell-less eggs and a reduction in bicarbonate ion secretion. Zn deficiency results in economic damages in the poultry industry, reduction or cessation of egg production and poor hatchability (Ogbuewu & Mbajiorgu, 2022).

The amount of a nutrient reaches the systemic circulation and is involved in metabolism after being ingested and absorbed is called bioavailability (Ammerman et al., 1995). Many authors have discussed Zn bioavailability and accumulation and there is a difference in absorption of organic and inorganic supplements (Suttle, 2010).The mineral zinc can be generated as sulfates, oxides, carbonates, or chlorides, while organic zinc is available as the metal ion complexes attached to organic molecules (Attia et al., 2013). Zinc methionine, zinc lysine and zinc threoninate are zinc amino acid chelates improving absorption and availability by reducing antagonism compared to the inorganic form, which may react with no absorbable compounds in the gastrointestinal tract (Neto et al., 2020).Generally, chelated organic elements are regarded as rare elements which compared to the mineral reserves, have a higher tissue availability (Mondal et al., 2009). Mabe et al. (2003) found the basic differences between the mineral and organic zinc reserves; however, regardless of the origin, the zinc, manganese, and copper supplements increase the eggshell breaking strength in aged laying hens. In addition, it has been suggested that the organically generated zinc has been prioritized in the cases in which the dietary levels of trace elements are pushed to the margins. Also, organic zinc (as the complex of metal amino acids) has been very effective in improving feed conversion ratio, broken eggs percentage, shell thickness, and Haugh Unit in the young laying hens.The findings of the meta-analysis of 11 research pieces indicate that dietary zinc supplementation bettered hen day egg production, feed conversion ratio, egg mass, egg weight, eggshell thickness, Haugh unit score, and blood zinc concentrations in laying hens while no significant effect on feed intake and eggshell weight was observed. In addition Subgroup analysis revealed a significantly higher hen daily egg production in laying hens fed a diet supplemented with zinc at 100 mg/kg feed. The subgroup findings indicate that specific parameters (zinc form, hen age at the start of the trial, supplementation duration, and inclusion level) were associated with the discrepancies observed between the experiments. Compared with the controls, the observed significant improvement in egg production and quality traits in laying hens fed Zn-supplemented diets will assist in sustainable use of zinc and policy advancements in the egg production industry. It is recommended to use the regression function in determining zinc inclusion levels for optimal quality and egg production in laying hens (Ogbuewu & Mbajiorgu, 2022).

Previous researchers (2008) observed no differences in bone quality and eggshell thickness when replacing inorganic zinc with zinc amino acid complexes (Swiatkiewicz and Koreleski, 2008; Gheisari et al., 2011). Various investigations have examined and compared the effects of different forms of mineral and organic zinc; however, the results are controversial in the scientific community regarding the origin of zinc supplementation when feeding laying hens .The disparity in the results of the investigations can be attributed to differences in the employed research models in terms of the strain of chicken used, hen age, experimental and basal diet composition,replacement amount, and the origin of the organic mineral used (Gheisari et al., 2011; Ogbuewu and Mbajiorgu, 2022). The natural presence of Zn in raw materials is not enough alone to satisfy the Zn requirements (Attia et al., 2013). On the other hand, dietary supplementation of organic Zn could alleviate the negative effects of age on egg quality characteristics in laying hens (Behjatian Esfahani et al.,2021). Previous studies have shown that organic Zn has more advantages than inorganic Zn in improving eggshell quality. However, the optimum supplementation amount has not been fully studied. The present study was designed to evaluate the effect of Zn supplementation on the antioxidant activity, bone strength, as well as zinc deposits in late-phase laying hens. Since there is no consensus on whether zinc supplementation can improve zinc deposition in the body tissues and compensate for the adverse effects of hen age on health, prompted us to investigate mineral deposition in the body tissues and bone health of Leghorn laying hens in their last laying phase when the bone mineral reserves would be significantly required.

1. Materials and methods

Study design

In the current study, 175 W36 Leghorn laying hens aged 80 weeks were chosen for the experiment. This experiment was performed in a commercial poultry house. The hens were kept in wire cages 52 cm long, 34 cm wide, and 30 cm high on three floors with a maximum temperature of 25°C, and a humidity of 56%, and a lighting period of 16 h. There were two nipple drinkers in each cage, and the feed was freely available to the hens. To conduct this research, the mineral supplement of zinc sulfate ($5H_2O$) with 34.12% pure zinc as the mineral base in the diet and Zinpro Company supplement with 12% pure zinc as an organic source in the diet were used.

The diets were adjusted based on the requirements recommended for the strain (Hyline International, 2020) and after reviewing the research conducted in this field according to Table (1). Zinc-free vitamins and mineral supplements were purchased.

Table 1. Diet Ingredients and Chemical Analysis

Ingredients	
Corn%	53.0
Soybean%	25.0
Calcium carbonate%	10.0
Vegetable oil%	1.40
Wheat bran%	6.08
Dicalcium phosphate%	2.20
Salt%	0.25
Sodium Bicarbonat %	0.15
[1]Vitamin supplement%	0.25
Mineral supplement%	0.25
DL-Methionine%	0.14
Lysine%	0.06
Chemical analysis	
Metabolizable energy (kcal / kg)	2810
Crude protein %	15.00
Calcium %	4.65
Available phosphorus %	0.4
Sodium %	0.18
Methionine%	0.32
Methionine-cysteine %	0.65
Lysine %	0.66
Arginine%	0.90
Threonine %	0.59

[1] Vitamin supplement composition for each kilogram of diet includes: 3200000 U/kg of vitamin A, 1320000 U/kg of vitamin D3, 8000 U/kg of vitamin E, 1000 mg / kg of vitamin K3, 1000 mg / Kg of vitamin B1, 2200 mg / kg of vitamin B2, 3200 mg / kg of vitamin Kalban, 12000 mg / kg of niacin, 1600 mg / kg of B6, 360 mg / kg of B9, 9 mg / kg of B12, 30 mg G / kg biotin, 44000 mg / kg choline,

antioxidant 3000 mg / kg. [2]Mineral supplement without Zn for each kilogram of basic diet includes: 79.01 mg of Mn, 3200 mg of copper, 480 mg of iodine, 88 mg of selenium, 16000 mg of iron. The amount of Zn in the basic diet measured in the laboratory was 63.58 mg / kg.

The birds fed the experimental diets in seven treatments with five replications. Each replication consisted of cages (five hens per cage), with 25 hens in each treatment. The hens under the experiment were first weighed, and those with the same weight were used for treatment to standardize the experiment conditions.In addition, for adaptation, the control diet was fed to the hens for two weeks. These diets, which contained different levels of mineral reserves and organic chelating agents, were prepared as presented in Table 2:

Table 2: Experimental Treatments[1]

Treatment	Zn sulfate	AA-Zn	Total zinc (mg/Kg)
	(%)		
T1	0	0	63.58
T2	15	0	78.58
T3	30	0	93.58
T4	45	0	108.58
T5	0	15	78.58
T6	0	30	93.58
T7	0	45	108.58

[1] Treatment 1: control (basal diet), treatment 2: basal diet with 15 mg/Kg of zinc sulfate, treatment 3: basal diet with 30 mg/Kg of zinc sulfate, treatment 4: basal diet with 45 mg/Kg of zinc sulfate, treatment 5: basal diet with 15 mg/Kg of organic zinc, treatment 6: basal diet with 30 mg/Kg of organic zinc, treatment 7: basal diet with 45 mg/Kg of organic zinc.

The amounts (percentage) of the zinc element provided in the sulfate zinc and organic zinc sources used for the experiment are shown in Table 3. Organic zinc supplement, as amino acid-Zn chelate (AA-Zn) with 12.5% Zn, was obtained from Zinpro Co.

Table 3. The Amounts of Provided Zinc Element in the Mineral and Organic Reserves Used for the experiment (%)

Treatment	**Zinc sulfate supplement**	**Organic zinc supplement**
Control (basal diet) *	0	0
a basal diet with 15 mg mineral zinc	4.53	0
a basal diet with 30 mg mineral zinc	9.06	0
a basal diet with 45 mg mineral zinc	13.6	0
a basal diet with 15 mg organic zinc	0	12.5
a basal diet with 30 mg organic zinc	0	25
a basal diet with 45 mg organic zinc	0	37.5

*the amount of basal diet analyzed in the laboratory was 63.58 mg/Kg. The Hy-Line strain need is 80 mg, according to the strain breeding guide (2016).

Production Performance:

Body weight was measured by weighing all hens individually at the onset and end of the experiment. Mortality was calculated during the experiment. Egg production was recorded daily from 80 to 92 weeks of age. All eggs were weighed by an electronic balance (Sartorius, accuracy 0.001). Feed intake was recorded weekly. The feed conversion ratio was calculated as the ratio of the mass of feed consumed per egg mass-produced. According to the following formula, egg mass production was determined based on the percent of hen-day production. Egg production was calculated based on hen house as shown in the formula below:

$$Average\ egg\ mass = \%hen - day\ production \times Average\ egg\ weight\ in\ gram$$

$$Hen\ house\ egg\ production = \frac{Total\ number\ of\ eggs\ laid\ during\ the\ period}{Total\ number\ of\ hens\ housed\ at\ the\ beginning\ of\ laying\ period}$$

Antioxidant Activities

To evaluate the antioxidant status, the activity of serum superoxide dismutase (SOD) enzyme was assessed using Randsel SOD diagnostic kit (Randox, Crumlin, Uk) in Mad laboratory. At the end of the experiment period, two hens were selected from each repetition to collect their blood samples (10 ml) during slaughtering.

Bone Strength

The bone strength samples were collected in end experiment. In brief, ten hens/treatments were selected and sacrificed by rapid cervical dislocation, and the right tibia was removed to analyze tibia breaking strength using an Instron device (Model H5KS, Tinius Olsen Company).

Zinc Deposition in Body Tissues

Two hens were selected from each replication to evaluate the level of zinc deposition in body tissues at the end of the experiment, and 10 mL of blood was collected from each hen during slaughter. Before slaughter, all experimental hens were fasted for 10 h to improve the accuracy of blood factor measurements. Serum samples were centrifuged at 3000 rpm for 10 min to separate the serum and were stored at -20°C until analysis. Moreover, to evaluate the amount of zinc in the bones, the tibia was dried in an oven for 24 h and then placed in an oven at 550°C to prepare the ash. The blood sample, tibia, yolk, and liver of two replication samples were selected and sent to the Laboratory to determine the zinc level. The zinc levels of the serum and other tissues were analyzed by Inductively coupled plasma-optical emission spectrometry (ICP-OES, Optima 8300, PerkinElmer, Waltham, MA, USA) according to the method suggested by Qiu et al. (2020).

Statistical Analysis

This experiment was conducted using simple Complete random design (Performance traits) for 12 weeks by using the following model.

$$Yij = \mu + Ti + eij$$

This experiments of SOD, bone strength, zinc deposition was conducted in the form of simple Complete random design with several observations in each repetition.The samples of SOD, bone strength, zinc deposition were collected in end experiment by using the following model.

$$Yijk = \mu + T_i + e_{ijk} + \varepsilon_{ijk}$$

where μ demonstrates the mean trait, Ti indicates the treatment effect, and ε ijk and eijk illustrate the effect of sampling and experimental error, respectively. The data were analyzed by using GLM in SAS statistical software.Tukey's test was utilized to compare the mean of various treatments when the mean difference was significant at the level of $P<0.05$.

2. Results and Discussion

Production performance

Performance of laying hens Results showed (Table 4) a significant difference in aged laying hens' production traits. Based on the obtained results, the feed intake, egg production (%), egg weight, egg mass, and conversion ratio were affected by the experimental treatments ($P < 0.05$). Results regarding the performance of laying hens showed are contrary to the findings of Stefanello et al. (2014), who reported that mineral trace elements in laying hens' diet had no effects on the feed intake, feed conversion ratio, and egg production.Manangi et al. (2015) also reported no effects for rare organic minerals. In a study on 96 Hy-Line W-36 Leghorn hens, the effects of adding zinc to the diet of laying hens on their performance and behavior were evaluated for ten weeks using four treatments (40, 70, 100, and 130 mg/kg zinc per diet). The results indicated that egg production in the hens receiving 130 mg/kg zinc was significantly higher than the hens receiving 40 and 70 mg/kg zinc ($P < 0.05$). Moreover, the hens receiving 130 mg/kg zinc had a lower feed intake than the other two treatments ($P < 0.05$).

Treatments 2, 4, 5, 6, and 7 demonstrated the highest feed intake. It has been declared that zinc influences appetite because anorexia is one of the symptoms of zinc deficiency (Behjatian Esfahani et al.,2021). The results of this study are in line with the results of Manangi et al. (2015) and Stefanello et al. (2014), that found no significant impacts of organic trace minerals (Mn, Zn, and Cu) and supplemental inorganic in laying hens on feed intake. Nevertheless, Gheisari et al. (2011) reported that the age of hens had a vital impact on the average daily feed intake.The findings demonstrated (in Table 4) a significant difference between Treatment 4 (45 mg/kg Zinc sulfate) and treatment 1 (basal diet) regarding egg production ($P < 0.05$). The cause of enhancement in egg production can be ascribed to the impact of zinc on egg production as a result of the impact of zinc in the albumin deposition in the formation of eggshell layers in the isthmus, the magnum, and eggshell formation in the uterus, and the impact of zinc on the enhancement in FSH, and LH hormones, progesterone, and estrogen (Park et al., 2004). Improvement in egg production is also probably due to the impact of zinc on the secretion of reproductive hormones. A considerable enhancement in the concentration was observed in the experiment on broiler estrogen and progesterone breeders by adding different zinc levels to the diet. Ovarian estrogen releases oviduct growth and enhances proteins, vitamins, blood calcium, fats, and nutrients required for egg formation.

In addition, progesterone induces mature follicles ovulation (Behjatian Esfahani et al., 2021). EL-Katcha et al. (2018) reported that adding 30 ppm of zinc nanoparticles to the diet increases egg production. Abedini et al. (2018) experimented on 288 Leghorn laying hens aged 64 weeks with four treatments. The experimental diets included a soybean meal (without Zn supplement) and a basal diet with 80 mg/kg ZnO, ZnO nanoparticles, and Zn-methionine. The results indicated that the egg production and mass in the treatments containing Zn-methionine and ZnO nanoparticles were significantly higher ($P < 0.05$). So, it is obvious that dietary supplementation of Zn sources could improve significantly egg production.

The mean egg weight at the end of the experimental period was significantly higher in Treatment 4 (45 mg/Kg of zinc sulfate),Treatment 5 (15 mg/kg chelated organic zinc), Treatment 6 (30 mg/kg chelated organic zinc), and Treatment 7 (45 mg/kg chelated organic zinc) ($P < 0.05$) than the control treatment and other treatments ($P > 0.05$). Li et al. (2019c) stated that the treatment containing 80 mg/kg zinc in the diet led to higher egg weight.It should be noticed (Table 4) that in all of the ages, treatments with 45 mg/kg sulfate zinc had the highest egg mass than inorganic Zn and the control group ($P < 0.05$) and ($P < 0.05$). This egg mass enhancement can be because of the effect of zinc in several physiological and biochemical processes. Zinc is needed to function normally in enzymes, protein structures, hormones, and development and growth. This improvement can result from zinc participation in the body's metabolism of energy and protein (Behjatian Esfahani et al., 2021). Because egg mass is related to egg production, the egg mass was bettered by enhancing egg production (Table 4) in hens fed the diet enriched with zinc sulfate and zinc organic than control.

The results of this study agreed with Behjatian Esfahani et al. (2021), who declared that egg mass increased considerably due to using organic zinc sources in the diet.Results (Table 4) demonstrated) that the lowest conversion ratio was observed in Treatments 4, 5, 6, and 7 ($P < 0.05$). Abd El-Hack et al. (2018) reported that Zn-Met improved the feed conversion ratio in laying hens. The addition of organic zinc to the diet also led to the improvement of the feed conversion ratio (P < 0.05) (Sorosh et al., 2018). In another study, Jialing et al. (2020) stated that the use of a diet with one-third of the commercial level of rare minerals (iron, copper, manganese, and zinc) reduced the number of daily eggs and egg mass and increased the ratio of feed intake to egg production compared to the other treatments (P < 0.05). Therefore, dietary organic Zn plays a vital role in increasing the digestion and absorption of nutrients in the gastrointestinal tract (Behjatian Esfahani et al.,2021), which improves the laying hens' performance in the present experiment.

Table 4. Comparison between the Increasing Levels of the Chelated Organic and Mineral Zinc in terms of Performance Traits of Aged Laying Hens

Treatment	Feed intake (g)	Hen day egg production (%)	Egg weight (g)	Egg mass (g)	Conversion ratio
1	107.6[b]	62.7[b]	65.4[b]	40.9[b]	2.62[a]
2	108.8[a]	65.5[ab]	66.4[ab]	43.4[ab]	2.51[ab]
3	108.7[ab]	64.6[ab]	66.6[ab]	43.1[ab]	2.52[ab]
4	109.3[a]	72.2[a]	67.5[a]	48.6[a]	2.26[b]
5	109.4[a]	70.1[ab]	67.9[a]	47.6[a]	2.32[b]
6	109.1[a]	70.1[ab]	68.4[a]	47.9[a]	2.30[b]
7	108.9[a]	68.7[ab]	68.4[a]	46.9[ab]	2.32[b]
SEM	0.274	0.484	0.388	1.42	0.065
P-value	<0.0009	<0.0151	<0.0001	<0.0022	<0.0012

[a,b] The means with dissimilar letters in each colum are statistically significantly different (P<0.05), SEM :Standard Error Mean, Treatment 1: control (basal diet), treatment 2: basal diet with 15 mg/Kg of zinc sulfate, treatment 3: basal diet with 30 mg/Kg of zinc sulfate, treatment 4: basal diet with 45 mg/Kg of zinc sulfate, treatment

5: basal diet with 15 mg/Kg of organic zinc, treatment 6: basal diet with 30 mg/Kg of organic zinc, treatment 7: basal diet with 45 mg/Kg of organic zinc.

Antioxidant status

The antioxidant effects of different Zn sources and doses on laying hens are reported in Table 5. According to the results of this experiment, SOD was significantly affected by the experimental treatments ($P < 0.05$). SOD, which is the primary antioxidant enzyme that protects cells from oxidative stress, increases the antioxidant ability to eliminate overactive reactive oxygen species (ROS) and reduce lipid peroxidation in poultry (Zhu et al., 2017). In a previous study, hens supplemented with Zn at a concentration of 70 mg/kg dry matter (DM) had higher Zn21 and Cu-ZnSOD (($P < 0.05$)content and tended to have higher serum concentrations of SOD than hens supplemented with Zn at the concentration of 35 mg/kg DM; however, the source of Zn did not have an effect (YU et al., 2020). One of the most important actions of Zn is the participation in the antioxidant defense system. Oxidative damage to the cell membrane has occurred through free radicals during zinc deficiency (Prasad and Kucuk ,2002).On the other hand, zinc is a cofactor of the main antioxidative enzyme Cu Zn-superoxide dismutase; it may play a key role in suppressing free radicals and in inhibiting NADPH-dependent lipid peroxidation (Prasad ,1997) as well as in preventing lipid peroxidation via inhibition of glutathione depletion (Gibbs et al. 1985). One of the proposed mechanisms of zinc's function is its capacity to replace transition metals (Fe, Cu) from binding sites. Zinc can compete with iron and copper to bind to the cell membrane and decrease the production of free radicals, thus exerting a direct antioxidant action (Tate et al. 1999). Zinc induces the production of metallothionein, an effective scavenger of hydroxyl radicals and it has been suggested that Zn-metallothionein complexes in the islet cells provide protection against immune-mediated free-radical attack (Shaheen and Abd El-Fattah ,1995).Zinc is an essential component of SOD, which has an important function in the detoxification of superoxide free radicals and the protection of cells against oxidative stress by catalyzing the conversion of superoxide anion (O_2) to H_2O_2 (YU et al., 2020). Zn has antioxidant properties and is essential for the immune system (Ogbuewu and Mbajiorgu, 2022).

Table 5: Effects of different Sources of organic and inorganic zinc on Antioxidant activity in old laying hens

Parameters	Experimental diets[1]							SEM	P-value
	T1	T2	T3	T4	T5	T6	T7		
SOD (U/ml)	221.2^{b}	285.7^{ab}	332.0^{a}	341.4^{a}	314.4^{a}	279.2^{ab}	306.6^{a}	0.694	0.0001

SEM: standard error meaning.

[a,b] Values in a row with no common superscript letter are significantly different ($P < 0.05$).

[1] Treatment 1: control (basal diet), treatment 2: basal diet with 15 mg/Kg of zinc sulfate, treatment 3: basal diet with 30 mg/Kg of zinc sulfate, treatment 4: basal diet with 45 mg/Kg of zinc sulfate, treatment 5: basal diet with 15 mg/Kg of organic zinc, treatment 6: basal diet with 30 mg/Kg of organic zinc, treatment 7: basal diet with 45 mg/Kg of organic zinc.

Bone Strength

The results regarding the effects of inorganic and organic zinc supplementation on tibia bone strength are presented in Table 6. According to these results, no significant differences were observed among the experimental treatments ($P > 0.05$). However, the cortical thickness of the bone with 30 g of chelated organic zinc (T6) was significantly increased compared to that in the control and the other experimental treatments.

Table 6: Effect of feeding supplemental inorganic and organic Zn on bone strength in old laying hens.

Treatment[1]	Force of break (N)	Work to break point (J)	Bend to break point (mm)	Cortical thickness (mm)
T1	212.3	223.3	0.146	2.00^{b}
T2	143.2	189.8	0.147	2.61^{b}
T3	193.8	200.8	0.124	3.10^{b}
T4	148.6	179.3	0.148	3.85^{b}
T5	229.1	177.5	0.109	4.29^{b}
T6	131.6	197.9	0.192	5.94^{a}
T7	212.3	224.8	0.138	4.82^{b}
SEM*	28.8	33.6	0.02	0.21
P-Value	0.0880	0.9040	0.3480	0.0001

[1] Treatment 1: control (basal diet), treatment 2: basal diet with 15 mg/Kg of zinc sulfate, treatment 3: basal diet with 30 mg/Kg of zinc sulfate, treatment 4: basal diet with 45 mg/Kg of zinc sulfate, treatment 5: basal diet with 15 mg/Kg of organic zinc, treatment 6: basal diet with 30 mg/Kg of organic zinc, treatment 7: basal diet with 45 mg/Kg of organic zinc.

Zinc deficiency (10 mg/kg) in young hens has a limiting effect on bone formation. In broilers, an increase in the zinc level up to 100 mg/kg led to a significant improvement in bone strength and lowered injury risks. Manangi et al. (2015) reported no significant association between chelated trace minerals (CTM) diet and the mineral source on this parameter. However, they showed that the use of CTM resulted in the strongest bone strength values. Substitution of inorganic Zn and Mn with amino acid complexes in birds aged up to 70 weeks (Stefanello et al., 2014) had no significant effect on the tibia bone, and the authors speculated that bone properties in older birds might be less sensitive to the availability of Zn and Mn in the diet than eggshells.

The decrease in eggshell quality of aged laying hens is one of the most critical issues in the poultry industry. Although the majority nutrition research on shell quality has focused on macronutrients such as calcium, phosphorus, and vitamin D_3, several enzymes associated with trace elements are calcified and involved in eggshell formation. These trace elements include zinc and manganese, which are responsible for producing carbonate and mucopolysaccharides as cofactors of metalloenzymes and play an important role in forming eggshells (Swiatkiewicz & Koreleski, 2008 and Park et al., 2004).

In terms of tibia strength, the treatment group receiving 30 mg/kg organic zinc showed a significant increase in the level of flexion and the thickness of the bone tissue ($P < 0.05$); therefore, it can be concluded that the hens in this treatment group had higher bone strength than the other treatments. Moreover, the results related to the shell thickness revealed that the eggshell thickness in the treatment group receiving 30 mg/kg organic zinc was increased. However, this difference was not significant, which can be due to the different bioavailability of zinc in the shell and bone. The antagonist relationships between zinc and calcium and how the carbonic enzyme anhydrase works in response to these two elements must be investigated.

Cufadar et al. (2020) stated that different levels of zinc and their interactions had no effects on the tibia weight, tibia pressure, and tibia breaking strength as the mechanical parameters of the tibia ($P > 0.05$). Replacement of zinc and manganese oxides with micronutrient amino acid complexes also did not affect the physical and geometric parameters of the tibia and ash content in the tibia and toes (Swiatkiewicz and Koreleski, 2008). Manangi et al. (2015) showed that mineral elements such as calcium, zinc, copper, and manganese led to the highest bone strength, although not statistically significant ($P = 0.01$). These results indicate other advantages of using organic sources of zinc over inorganic sources.

Nevertheless, adding Zn has been declared to enhance Ca's utilization in hens and better the eggshell's qualitative parameters (Klecker et al. 2002). It has also been reported that zinc supplementation increases eggshell quality since it is an element of the carbonic anhydrase enzyme, supplying the carbonate ions during eggshell formation (Ogbuewu & Mbajiorgu, 2022). Moreover, it has been indicated that the secretion of alkaline phosphatase is decreased under stress conditions, and this enzyme interacts with zinc in the bone's calcium storage (Rajabi & Torki, 2021).

Zinc Deposition in Body Tissues

Table 7 presents the zinc deposition in the aged laying hens' body tissues. The zinc deposition was remarkably different between the yolk, tibia, liver, and serum treatments ($P < 0.05$).

Abedini et al. (2018) investigated three sources of zinc, namely zinc oxide, zinc oxide nanoparticles, methionine-zinc, and reported that zinc deposition in the tibia, liver, pancreas, eggs, and feces was significantly different among the treatments ($P < 0.01$). Zinc has a protective effect against oxidative damage in pancreatic tissue; and it may help the pancreas function properly, including secretion of digestive enzymes, thus improving the digestibility of nutrients (Onderci et al., 2003).

Table 7. Zinc Deposition in the Body Tissues of Aged Laying Hens

	Bone (tibia)	**Yolk**	**Egg shell**	**Serum***	**Liver**
T1	173.4^{b}	17.4^{b}	0.56	6.47^{b}	83.5^{b}
T2	213.9^{a}	29.5^{ab}	0.57	6.94^{ab}	111.5^{a}
T3	216.5^{a}	40.6^{a}	0.63	7.34^{a}	81.6^{ab}
T4	205.1^{a}	54.2^{a}	0.56	5.38^{b}	137.9^{a}
T5	236.2^{a}	46.2^{a}	0.61	6.93^{ab}	101.3^{ab}
T6	193.1^{a}	37.6^{a}	1.01	6.16^{ab}	74.4^{ab}
T7	210.1^{a}	65.3^{a}	1.18	7.36^{a}	64.2^{ab}

SEM	3.56	3.08	0.20	0.20	5.11
P-Value	0.088	0.040	0.348	0.0001	0.030

[*]per ppm/ml. [a.b] The means with dissimilar letters in each column are statistically significantly different (P<0.05), SEM: Standard Error of Mean

[1] Treatment 1: control (basal diet), treatment 2: basal diet with 15 mg/Kg of zinc sulfate, treatment 3: basal diet with 30 mg/Kg of zinc sulfate, treatment 4: basal diet with 45 mg/Kg of zinc sulfate, treatment 5: basal diet with 15 mg/Kg of organic zinc, treatment 6: basal diet with 30 mg/Kg of organic zinc, treatment 7: basal diet with 45 mg/Kg of organic zinc

Bone (Tibia) Deposition

The measurement of zinc bioavailability in the right tibia showed that this parameter was significantly higher in treatments containing sulfate Bone (tibia) deposition and organic zinc than in the control treatment ($P < 0.05$).Treatment 5 (containing 15 mg/kg organic zinc) had the highest zinc concentration with 236.23 ppm. Cufadar et al. (2020) examined three sources of zinc (zinc oxide, zinc proteins, and zinc oxide nanoparticles) at 20, 40, 60, 80, and 100 mg/kg levels in laying hens' diet. They indicated that different sources of zinc had a significant effect on the zinc levels, although did not significantly affect the calcium and phosphorus content of the tibia ($P > 0.05$). Li et al. (2019b), in a study on 576 Lingnan parent broiler hens aged 58 weeks, reported that the hens receiving 24 to 48 mg/kg zinc had a higher tibia breaking strength than those in the other experimental treatments. In addition, they reported that the increase in Zn-Met-induced calcium deposition might be due to the increase in zinc content in the serum and tissues, which is attributed to the rise in serum calcium and albumin concentrations and the concentration of calcium, albumin, and CaBP-D28k mRNA in the eggshell gland (ESG) (Li et al. (2019a).

Yolk Deposition

In addition, the zinc bioavailability level in the yolk in sulfate and organic zinc supplement treatments had a significant increase compared to the control treatment, except for Treatment 2 ($P < 0.05$). Li et al. (2019a) compared three levels of 0, 40, and 80 mg of zinc sulfate and methionine-zinc. They stated that the treatment containing 80 mg/kg methionine-zinc led to a higher calcium deposition on the transverse section of the eggshell and had a higher zinc content in the yolk and serum ($P < 0.05$).

Saleh et al. (2020) evaluated laying hens receiving a mixture of 0.5 and 1 g of sulfate and organic zinc, manganese, and copper compared to the control treatment and reported that the yolk's zinc and copper levels were significantly increased compared to the control treatment ($P < 0.05$), while manganese was only numerically increased ($P > 0.05$).

Shell Deposition

Based on the obtained results, there were no significant differences between the control treatment and the treatments containing sulfate and organic zinc supplements in terms of the shell's zinc levels ($P > 0.05$). However, hens receiving 30 and 45 mg/kg organic zinc treatments had the highest numerical values of zinc in their shells.

Serum Deposition

The treatment containing 45 mg/kg sulfate zinc (Treatment 4) led to a significant decrease in serum zinc compared to the control treatment (5.38 ppm). The treatment containing 45 mg/kg organic zinc (Treatment 3,7) led to a significant statistical increase compared to the control treatment (7.36 ppm). Abd El-Hack et al. (2018) reported that the ZnO or Zn-Met supplement increased serum zinc compared to the control group, and there were no differences between the supplemental doses of zinc. In the current study, serum zinc was enhanced in the treatment group receiving 30 mg/kg zinc sulfate and 45 mg/kg organic zinc, while there was no increase in the other treatments.

Liver Deposition

The zinc bioavailability in the liver showed a significant difference since Treatment 2 (15 mg/kg zinc sulfate) and Treatment 4 (45 mg/kg zinc sulfate), had the highest zinc levels with 111.51 and 137.99 ppm, respectively ($P < 0.05$). Diet zinc sources significantly affected the liver zinc levels ($P < 0.05$). The zinc deposition in body tissues is a part of the digested and absorbed elements used in animal metabolism. Therefore, selecting a precise criterion is of great importance in estimating bioavailability. The bioavailability of different mineral combinations changes when added to organic or inorganic carriers (Ammerman et al., 1995). The bioavailability changes with the alterations of the absorption mechanism, e.g., transmission through the cell wall by diffusion are much less efficient than the translocation of amino acid-bonded minerals.

Furthermore, it should be noted that there are some special interactions between high-consumption and low-consumption mineral elements that can lead to antagonism or cooperation between them. Organic Zn sources are more biologically available in the digestive system than inorganic sources, resulting in lesser Zn excretion (Ogbuewu & Mbajiorgu, 2022). However, similar results regarding the zinc reserves in body tissues both in sulfate and organic forms of zinc. Qifang Yu et al. (2020) studied the effects of two organic sources of amino acid-zinc and zinc sulfate at two concentrations of 35 and 70 mg/kg. They reported that the zinc content of the eggs was linearly increased for both sources.

Bioavailability can be specified as the degree to which an ingested nutrient in a specific source is absorbed in a form that the animal can utilize in metabolism (Ammerman et al., 1995). Relative Bioavailability Value (RBV), or the bioavailability of a nutrient within one source, is evaluated concerning the bioavailability of the same nutrient in a standard source. Evaluating the minerals' storage or deposition into chosen tissues is the utmost common variable evaluated in trace mineral RBV experiments (Richards et al.,2015). Zinc in chelate creates a soluble complex with zinc, and as a result, it is readily available to the animal. An alternative possibility is that organic Zn sources are absorbed through amino acid or peptide transport systems, leading to higher bioavailability and digestibility (Behjatian Esfahani et al., 2021). Consequently, higher bioavailability Zn probably has a significant role in enhancing the absorption of nutrients and digestion, improving the egg quality and performance of laying hens in the present experiment.

In previous studies,the addition of 30 ppm of Zn oxide also increased the level of Zn in the excreta (Attia et al.,2013).Tsai et al (2016), concluded that 60 mg/kg nanosized zinc oxide can enhance zinc absorption and showed its feeding effective in aged layers. Moreover, the recommendation level of MHA-Zn in the diet of aged laying hens is 40 mg/kg (Min et al.,2018).The highest level of supplementation (120 ppm) was correlated to a high zinc excretion for both zinc sulfate and organic zinc. This indicates that the hens are not retaining a large portion of the zinc that is fed. Overall, the zinc content of the manure decreased for all treatments during the trial period. This could indicate that, as the hen's body adapts to the diet, it can better utilize the nutrients in the feed and therefore, show decreased manure zinc excretion.

The linear regression analysis showed a strong linear relationship between zinc level in the diet and feed intake and egg weight and poly relationship between zinc level and egg production and egg mass. The prediction equations will be valuable to producers who may be interested in feeding higher levels of zinc for the proposed benefits, but also want to monitor zinc levels in the manure. This will provide them with an accurate indicator of zinc excretion in the manure based on the zinc source and content of the proposed diet fed.In this experiment,the linear regression analysis of zinc source showed a a strong linear relationship between zinc source and performance traits. The level of 30 mg/kg organic zinc and 45 mg/Kg of zinc sulfate supplementation was high correlated with performance traits.According to my findings, organic Zn was more efficient that inorganic Zn in improving bone strength of laying hens.

Conclusion

This study's results the denotes the importance of supplementary dietary Zn from different sources that improve feed intake and antioxidant activity in older laying hens. Zn supplements increased egg weight, and organic Zn supplement boosted egg mass and decreased Feed Conversion Ratio. The Zn deposition was raised in the tibia and egg yolk, liver by Zn supplementation. As it can be concluded, dietary supplementation of Zn could alleviate the negative effects of age on antioxidant status and bone characteristics in laying hens. The optimum supplementation amount was of 30 mg/kg organic zinc in the late phase of laying improved the cortical thickness of the bone in older laying hens.

Literature Cited :

1. Abd El-Hack, Mohamed E., Mahmoud Alagawany, Ayman S. Salah, Mervat A. Abdel-Latif & Mohamed F. A. Farghly. (2018). Effects of Dietary Supplementation of Zinc Oxide and Zinc Methionine on Layer Performance, Egg Quality, and Blood Serum Indices. Biological Trace Element Research. 184, pages 456–462.
2. Abedini, M., F. Shariatmadari, M. A. Karimi Torshizi, H. Ahmadi (2018): Effects of zinc oxide nanoparticles on the egg quality, immune response, zinc retention, and blood parameters of laying hens in the late production phase. Journal of Animal Physiology and Animal Nutrition.
3. Ammerman, C. B., D. H. Baker, A. J. Lewis (1995): Bioavailability of Nutrients for Animals. AminoAcids, Minerals and Vitamins.Academic Press, San Diego, CA.
4. Attia, Y.A., A. Al-Hamid, H. S. Zeweil, E. M. Qota, F. Bovera, G. Monastra, M. D. Sahledom (2013): Effect of Dietary Amounts of Inorganic and Organic Zinc on Productive and Physiological Traits of White Peking Ducks. Animal. 7(6), 895-900.
5. Behjatian Esfahani, M., H. Moravej,M. Ghaffarzadeh, and G. A. Nehzati Paghaleh.(2021). Comparison the Zn-threonine, Zn-methionine, and Zn oxide

on performance, egg quality, Zn bioavailability, and Zn content in egg and excreta of laying hens. Biological trace element research, 199(1), 292-304.

6. Cufadar, Y., G. Rabia, K. Gülşah, Y. Bedia (2020): Effects of Dietary Different Levels of Nano, Organic and Inorganic Zinc Sources on Performance, Eggshell Quality, Bone Mechanical Parameters and Mineral Contents of the Tibia, Liver, Serum and Excreta in Laying Hens. Biological Trace Element Research. 193, 241–251.
7. El-katcha, M. I., M. A. Soltan, M. M. Arafa, k. El-Naggar, E. S. R. Kawarei (2018): Impact of Dietary Replacement of Inorganic Zinc by Organic or Nano Sources on Productive Performance, Immune Response and Some Blood Biochemical Constituents of Laying Hen. Alexandria Journal of Veterinary Sciences 59(1), 48-59. DOI: 10.5455/ajvs.301885.
8. Gheisari, A. A., A. Sanei, A. Samie, M. M. Gheisari, M. Toghyani (2011): Effect of Diets Supplemented with Different Levels of Manganese, Zinc and Copper from their Organic or Inorganic Sources on Egg Production and Quality Characteristics in Laying Hens. Biological Trace Element Research. 142, 557-571.
9. Gibbs, P. N., M. Gore, and P. M. Jordan (1985). Investigation of the effect of metal ions on the reactivity of thiol groups in human 5-aminolaevulinate dehydratase. Biochemical Journal, 225(3), 573-580.
10. Jialing, Q., Xintao, L., Lianxiang, M., Chuanchuan, H., Junna, H., Bing, L., Dongyou, Y., Gang, L., and Jiming, X. (2020). Low-dose of organic trace minerals reduced fecal mineral excretion without compromising performance of laying hens. Asian-Australasian Journal of Animal Sciences, 33(4): 588-596.
11. Klecker, D., Zeman, L., Jelinek, P., & Bunesova, A. (2002). Effect of manganese and zinc chelates on the quality of eggs. Acta Universitatis Agriculturae et Sylviculturae Mendelianae Brunensis (Czech Republic).
12. Li, L., M. Liping, Zh. Mingkun, W. Liansong (2019a): Dietary addition of zinc-methionine influenced eggshell quality by affecting calcium deposition in eggshell formation of laying hens. British Journal of Nutrition. 122(9), 961-973.

13.Li. L., K. F. M. Abouelezz, Z. Gou, X. Lin, Y. Wang, Q. Fan, Z. Cheng, F. Ding, S. Jiang, Z. Jiang (2019b): Optimization of Dietary Zinc Requirement for Broiler Breeder Hens of Chinese Yellow-Feathered Chicken. Animals. 9(472), 1-14. doi:10.3390/ani9070472

14.Li. L.L., Y.J. Gong, Q. ZhanH, Y. X. Zheng, X. T. Zou (2019c): Effects of dietary Zn-methionine supplementation on the laying performance, egg quality, antioxidant capacity, and serum parameters of laying hens. Poultry Science. 98(2), 923-931.

15.Min, Y. N., Liu, F. X., Qi, X., Ji, S., Ma, S. X., Liu, X., Z. P. Wang,and Gao, A. Y. (2018). Effects of methionine hydroxyl analog chelated zinc on laying performance, eggshell quality, eggshell mineral deposition, and activities of Zn-containing enzymes in aged laying hens. Poultry science, 97(10), 3587-3593.

16.Mabe, I., C. Rapp, M. M. Bain, Y. Nys (2003): supplementation of a corn-soybean meal diet with manganese, copper, and zinc from organic or inorganic sources improves eggshell quality in aged laying hens. poultry science. 82, 1903-1913.

17.Manangi, M. K., M. Vazques-Anon, J. D. Richards, S. Carter, and C. D. Knight (2015): The impact of feeding supplemental chelated trace minerals on shell quality , tibia breaking strength, and immune response in laying hens. J. Appl. Poult. Res. 24, 316–326.

18.Mondal, S., S. Haldar, P. Saha, T. K. Ghosh (2009): Metabolism and Tissue Distribution of Trace Elements in Broiler Chickens' Fed Diets Containing Deficient and Plethoric Levels of Copper, Manganese, and Zinc. Biological Trace Element Research. 137, 190-205.

19.Neto, M.T., Dadalt, J.C. and M.L.P.Tse (2020). Dietary combination of chelated zinc and threonine and effects on egg production, egg quality and nutrient balance of Brown laying hens from 20 to 49 weeks of age. Animal Feed Science and Technology.267, p114555.

20.Ogbuewu, I. P., and C. A. Mbajiorgu (2022). Meta-analysis of Zinc Supplementation on Laying Performance, Egg Quality Characteristics, and

Blood Zinc Concentrations in Laying Hens. Biological Trace Element Research, 1-17.

21. Onderci, M., N., Sahin, K., Sahin and N. Kilic (2003). Antioxidant properties of chromium and zinc. Biological trace element research, 92(2), 139-149.
22. Prasad, A. S. (1997). The role of zinc in brain and nerve functions. In Metals and oxidative damage in neurological disorders (pp. 95-111). Springer, Boston, MA.
23. Prasad, A. S., and O. Kucuk (2002). Zinc in cancer prevention. Cancer and metastasis Reviews, 21(3), 291-295.
24. Park, S. Y., S. G. Birkhold, L. F. Kubena, D. J. Nisbet, S. C. Ricke (2004): Review on the role ofdietary zinc in poultry nutrition, immunity and reproduction. Biological Trace Element Research, 101(2), 147-163.
25. Qifang, Yu., L. Hu, Y. Kaili, T. Xiaopeng, Ch. Sijia, M. A. Kolapo, D. Allan, F. Rejun (2020): Effect of the level and source of supplementary dietary zinc on egg production, quality, and zinc content and on serum antioxidant parameters and zinc concentration in laying hens. Poultry Science. 99(11), 6233-6238.
26. Qiu, J., X. Lu, Ma, C. Hou, J.He, B. Liu, D. Yu, G. Lin and J. Xu (2020). Low-dose of organic trace minerals reduced fecal mineral excretion without compromising performance of laying hens. Asian-Australasian Journal of Animal Sciences, 33(4), 588.
27. Rajabi, M., and M.Torki (2021). Effect of dietary supplemental vitamin C and zinc sulfate on productive performance, egg quality traits and blood parameters of laying hens reared under cold stress condition. Journal of Applied Animal Research, 49(1), 309-317.
28. Richards, J. D., P. M. Fisher, J. L. Evans and K. J. Wedekind (2015). Greater bioavailability of chelated compared with inorganic zinc in broiler chicks in the presence or absence of elevated calcium and phosphorus. Open Access Animal Physiology, 7, 97.
29. Saleh, A. A., M. S. Eltantawy, E. M. Gawish, H. H. Younis, Kh. A. Amber, A. E. M. E. Abd El-Moneim, T. A. Ebeid (2020): Impact of Dietary Organic

Mineral Supplementation on Reproductive Performance, Egg Quality Characteristics, Lipid Oxidation, Ovarian Follicular Development, and Immune Response in Laying Hens Under High Ambient Temperature. Biological Trace Element Research. 195, 506–514.

30. Shaheen, A. A., and A. A. Abd El-Fattah (1995). Effect of dietary zinc on lipid peroxidation, glutathione, protein thiols levels and superoxide dismutase activity in rat tissues. The international journal of biochemistry & cell biology, 27(1), 89-95.

31. Sorosh, Z., S. Salari, M. Sari, J. Fayazi, S. Tabatabaei (2018): Dietary zinc supplementation and the performance and behaviour of caged laying hens. Animal Production Science 59(2), 331-337.

32. Stefanello, C., T. C. Santos, A. E. Murakami, E. N. Martins, T. C. Carneiro (2014): Productive performance, eggshell quality, and eggshell ultrastructure of laying hens fed diets supplemented with organic trace minerals. Poult. Sci. 93, 104–113. Available at http://www.ncbi.nlm.nih.gov/pubmed/24570429.

33. Suttle, N. F (2010): Zinc. In: Suttle, N.F., Mineral Nutrition of Livestock. 4th edition. Oxfordshire, UK: CAB International. 426-458.

34. Swiatkiewicz, S., J. Koreleski (2008): The effect of zinc and manganese source in the diet for laying hens on egg shell and bones quality. Veterinarni Medicina, 53(10), 555-563.

35. Tate Jr, D. J., M. V., Miceli and D. A. Newsome (1999). Zinc protects against oxidative damage in cultured human retinal pigment epithelial cells. Free Radical Biology and Medicine, 26(5-6), 704-713.

36. Tsai,Y.H.,S.Y.,Mao,M.Z.,Li, J.T. Huang,T.L.Lien (2016): Effects of nanosize zinc oxide on zinc retention, eggshell quality, immune response and serum parameters of aged laying hens. Animal Feed Science and Technology, 213,99-107.

37. Vakili, R., and A.A.Rashidi (2011). The effects of dietary fat, vitamin E and zinc supplementation on fatty acid composition and oxidative stability of

muscle thigh in broilers under heat stress. African Journal Agriculture Research, 6(12):2800–2806.

38. Yu, Q., Liu, H., Yang, K., Tang, X., Chen, S., Ajuwon, K.M., Degen, A. and Fang, R., (2020). Effect of the level and source of supplementary dietary zinc on egg production, quality, and zinc content and on serum antioxidant parameters and zinc concentration in laying hens. Poultry Science, 99(11),6233-6238.

39. Zhu, Y. W., Li, W. X., Lu, L., Zhang, L. Y., Ji, C., Lin, X., C. Liu, J. Odle and Luo, X. G., (2017). Impact of maternal heat stress in conjunction with dietary zinc supplementation on hatchability, embryonic development, and growth performance in offspring broilers. Poultry Science, 96(7), 2351-2359.

CONCLUSION

Zinc plays multiple roles in poultry nutrition and helps in increasing body weight, feed intake and feed efficiency in broilers. Zinc also provides better eggshell quality in layer birds. Its role in immunity and disease resistance is remarkable. Even though zinc as a nutrient is known over many decades, still there is a need for study in aspects of both nutrition and production performance. According to NRC (1994), usually zinc in poultry ration is not sufficient but often diets are formulated with zinc at the level of 120 mg/kg of diet. So, wide safety margins are often used to avoid trace mineral deficiencies and allow birds to achieve their genetic potential. This higher level of addition possibly reduces deficiency under commercial condition.

This study's results imply the importance of supplementary dietary Zn from different sources that improve feed intake in older laying hens. The use of 30 mg/kg Zinc methionine chelate zinc in the last phase of laying improved the cortical thickness of the bone in older laying hens. Zn supplements increased egg weight, and organic Zn supplements boosted egg mass and decreased Feed Conversion Ratio. The Zn deposition was raised in the tibia and egg yolk, liver by Zn supplementation. On the other hand, dietary supplementation of Zn could alleviate the negative effects of age on bone characteristics in laying hens. It is suggested that the additional expenses of purchasing supplementations be thoroughly investigated.

Effects of Zinc supplementation with dietary Fat, vitamin E and on tibia breaking strength in Female broilers under heat stress

Zinc is used in poultry diets because of its antistress effects. Moreover, its requirement increases and its retention decreases during stress (Bartlett and Smith, 2003; Sahin and Kucuk, 2003b).The result of Sandoval et al.(1998) who supplemented basal ration with zink amount between 1,000 and 1,500 mg/kg , showed that dietary zinc supplementation lead to increased tibia zinc concentration. Zinc deficiency also causes shortness and thickness in long bones of legs and wings. There is also some evidence suggesting supplemental Zinc can alleviate negative effects of heat stress in broiler chickens (Klasing, 1984).

Bone breaking strength is measured by evaluating the reaction of the bone to stress and force. An increase in bone mineralization is accompanied by an increase in bone stress and bending moment values. Bone mechanical strength is affected by nutrition, genes, age, sex, environment and etc., for the expression of collagen and proteins, quantity and quality of the organic and inorganic material, content and size of the mineral material, and design and structure of the bone (Boskey *et al.*, 1999). Velleman (2000) explained the structural complexity and composition of bone associated with strength, which varied according to the age and nutritional status of the bird:femur compression strength = 11.9 + 0.0141 (bird BW). However, contradictory data regarding the effect of vitamin E on collagen synthesis exists, and few studies have reported its influence on cartilage development. In fact, increasing in thickness of the growth plate was proportional to DL-α-tocopheryl acetate dose. The effect of vitamin E on chondrocytes was observed by Xu *et al.* (1995) in chicks fed by two levels of DL-α-tocopheryl acetate and two dietary lipids. In their study, the thickness of the entire growth plate cartilage and of the lower hypertrophic chondrocyte (mineralized) zone was significantly wider in animals fed the greater level of vitamin E. α-tocopherol protects PUFA from lipid oxidation. However, the effect of α-tocopherol and zinc supplementation on bone strengthen in poultry has rarely been studied and is rather controversial.

In this study, the relation between dietary fat, zinc and vitamin E with heat stress on femur breaking strength has been studied separately.Therefore, the objective of this experiment was to determine the effect of four mention object in performance parameters and TBS too.

Material and method

Experimental design, animals, housing and diets

Two hundred forty female Arbor Acres broiler chicks were used for evaluating effects of diet and temperature on the immune response and blood parameters. Two factors of temperature and feed (2×3) were considered in a complete random design to form 6 separate treatments. Each treatment had 4 replications and 10 birds and it lasted for 6 weeks. Diets included basal diet supplemented with %5 saturated fatty acid or unsaturated fatty and antioxidant (100 IU vitamin E and 50 mg/kg zinc); that birds received this 3 diet in two temperature (22°C and 32°C). Table 1 shows the ingredients and chemical analyses of the starter and grower diets fed to experimental broiler chickens. The chicks were placed in floor pens equipped with stainless steel feeders, and automatic water drinkers.Lighting was continuous and water and feed were available *ad libitum* for 56 th d. The weights of birds were recorded at the end of the experiment and also at 21 th d of age. Feed intakes per pen were recorded. Birds were always deprived of feed 12 h before being weighed. The experimental chickens were slaughtered at the 56th d of age.

Table 1. Ingredients and chemical analyses of the starter and grower diets.

	Starter percentage of DM			Grower percentage of DM		
treatment	O	A	T	O	A	T
Ingredients						
Ground corn	54.75	54.75	54	61.73	61.73	61
Soybean meal	37.28	37.28	37.28	30.25	30.25	30.25
Fish oil	4.51	4.51	-	4.65	4.65	-
Tallow	-	-	5.26	-	-	5.38
Dicalcium phosphate	1.37	1.37	1.37	1.33	1.33	1.33
Sodium chloride	0.1	0.1	0.1	0.2	0.2	0.02
Limestone ground	0.53	0.53	0.53	0.53	0.53	0.53
Calcium carbonate	0.1	0.1	0.1	0.16	0.16	0.16
DL-methionine	0.24	0.24	0.24	0.25	0.25	0.25
Lysine	-	-	-	0.12	0.12	0.12
Cavilamycin	0.1	0.1	0.1	0.1	0.1	0.1
Ca- propionate	0.1	0.1	0.1	0.1	0.1	0.1
Vitamin E IU	-	100	-	-	100	-

Zinc mg/Kg	-	50	-	-	50	-
Clinao	0.1	0.1	0.1	0.1	0.1	0.1
Toxin	0.3	0.3	0.3	-	-	-
Vitamin premix*	0.27	0.27	0.27	0.27	0.27	0.27
Trace mineral premix**	0.16	0.16	0.16	0.16	0.16	0.16
DM%	89.65	89.65	89.63	89.68	89.68	89.68
Chemical analyses						
ME, kcal/kg	3100	3100	3100	3200	3200	3200
Cp	23	23	23	20.03	20.03	20.03
Crude fat	7.04	7.04	7.04	6.35	6.35	6.35

* premix supplied for 2 kilogram: vitamin A, 15000 IU; cholecalciferol, 3 IU, vitamin E 15 IU; menadione, 2.5 mg; vitamin B1, 1 mg vitamin B2, 10mg; niacin, 70mg; d-pantotheenic acid, 20 mg; vitamin B12, 4 mg; folic acid, 2 mg; biotin, 0.1 mg.

** premix supplied for 2 kg; Mn, 80 mg: Mn, 80 mg; Fe, 25 mg; Zn, 50 mg; Cu, 7 mg; Iodine, 0.3 mg; Se, 0.15 mg; Choline chloride, 350 mg.

Bone sampling

At the end of the experiment (56th d), birds were killed by cervical dislocation and weighed, afterward the right tibia bones were removed. Much of the adhering tissues were removed from each bone at the time of dissection. Each bone was subsequently placed in boiling water for 15 min to complete tissue removal. The length of each bone was determined with a ruler, and the width was measured using a micrometer. Tibia

Bones from 10 birds of each group were evaluated. Tibia breaking strength was determined with instron[1] set. A compression test determined the behavior of materials under crushing loads. The biomechanical strength of each bone was measured using a material testing machine. The bones were held in vertical position and the compression load was applied in the cross-sectional area up to bone failure; the total crushing load at failure was determined in Newton (N/mm^2). The bone ash was determined after ashing in a muffle furnace at 600 °C for 24 h.

Bone strength

The effect of temperatures and diets are shown in table 2. There are significant differences of bone parameters in different groups. The highest weight of dry bone, length, width, ash and maximum force for breaking bone were observed in low environmental temperature breeding groups ($P<$ 0.05). High temperature causes excretion of some minerals for example Ca, Fe, Zn, and leads to decrease bone strengthen (Post *et al.,* 2003). In chickens, bone mineralization as determined through dual-energy x-ray absorptiometry (DEXA) is positively correlated with bone breaking force (r = 0.58 to 0.68; $P < 0.001$) and bone ash weight (r = 0.73 to 0.99; $P < 0.001$; Onyango *et al.,* 2003; Mazzuco and Hester, 2005; Schreiweis *et al.,* 2005). Higher increases in stress occur with smaller increases in ash content. Decrease in bone mineralization is accompanied by a decrease in bone breaking force and bone ash weight (Crenshaw *et al.*, 1981). Our results showed that oil supplementation can affect the TBS and weight of dry and wet bone ($P<$ 0.05). This result is in agreement with Liu *et al.* (2003) who showed high n-3 fatty acids increased the strength of bone in quail. In their experiment quail fed soybean oil diets (high in n-6) had significantly lower values in shear force and stress than those fed fish oil or hydrogenated soybean oil. Sakaguchi *et al* (1994) shown that dietary EPA and DHA led to an increase bone strength. It is unknown whether the potential effects of n-3 fatty acids on bone are due to specific longer chain derivatives, such as DHA and EPA or not. Thus, the differences

---

[1]Huns field model H 50 K-S

in specific n-3 sources on bone have to be clarified. The highest dry bone weight, width and length of bone, maximum force for breaking bone and ash value obtained in the group fed by diet supplemented with zinc and vitamin E. In fact, the increase in thickness of the growth plate was proportional to DL-α-tocopheryl acetate dose. The effect of vitamin E on chondrocytes was observed by Xu *et al.* (1995) in chicks fed two levels of DL-α-tocopheryl acetate and two dietary lipids. In their study, the thickness of the entire growth plate cartilage and of the lower hypertrophic chondrocyte (mineralized) zone was significantly wider in animals fed the greater level of vitamin E. The authors maintained that the increased thickness of the mineralized zone may be due to decreased cartilage resorption and phagocytic activity on the metaphyseal side. The lowest bone strengthen was observed in the group fed by diet supplemented with tallow. It seems that, high level of arachidonic acid in supplemented dietary with tallow, which is the precursor of PGE2 and may lead to decrease in bone formation, consequently decrease in bone strengthen (Watkins *et al.* ,1997). Our results are compatible with the results of Liu et al (2003). The highest bones strengthen, dry matter and ash content were observed in broilers fed by supplemented diet with zinc and vitamin E in normal temperature. Afterward the highest dry bone weight was observed in broiler under heat stress, but fed by diet supplemented by tallow. Higher growing rate in this group could be the reason of this fact. Feeding diet supplemented with zinc and vitamin E and diet supplemented with canola and fish oil in normal temperature resulted to maximum bone strength and ash content in broiler, respectively. It seems this high bone strength is because of high level of EPA, DHA, omega-3 and Zn in diet, which might increase Ca precipitation and collage formation. The lowest ash value and force for breaking bone were observed in broiler under heat stress, which fed by tallow. That would probably be due to high environmental temperatures and high omega-6 content in the diet. It might increase excretion some minerals and PGE formation, but decrease in collagen formation.

Table 2. Tibia lengths, widths, ash, breaking strength in broiler

Treatment	ash	Maximum force for breaking bones	Length of bones	Width of bones	Weight of Dry bones
	%	N/mm2	ml	mm	Gr
Interaction					
Temperatures					
Heat stress (H)	41.91 b	79.88 b	93.91 b	6.89 b	7.22 b
Normal (c)	46.10 a	103.67 a	103.40 a	7.85 a	10.15 a
diets					
O	42.7 b	86.92 b	96.75 b	7.22 b	8.51 b
T	41.15 b	88.37 c	93 c	6..9 b	8.71 c
A	48.2 a	110 a	106 a	8.3 a	9.48 a
Interaction(Temp × Feed)					
TH	40.16 d	66.25 e	96 bc	6.91 c	7.36 d
TC	42.77 c	90.50 c	100 b	7.46 b	10.22 b
OH	14.18 c	71.50 d	89.25 c	6.20 d	6.10 e
OC	44.32 b	102.35 b	98.75 b	7.09 c	9.20 c

AH	44.58 b	101.90 b	100.75b	7.06 c	7.95 d
AC	53.83 a	118.18 a	111.75 a	9	11.01 a
SEM	0.82	3.81	1.60	0.16	0.33

a b c d e , values that do not have common superscripts are significantly different (P <0.05).

O: Unsaturated fatty acid, A: basal diet + 100 IU/ kg vitamin E and 50 mg/kg zinc T: saturated fatty acid, OH: high temperature and O diet, OC: normal temperature and O diet, AH: high temperature and A diet, AC: normal temperature and A diet. TH: high temperature and T diet TC: normal temperature and T diet.

Conclusion

In conclusion, the higher environmental temperature and supplementing diet with zinc and vitamin E fish oil, lead to increase in bone strength in broiler.

Literature Cited :

Bartlett, J. R., and M. O. Smith. (2003). Effects of different levels of zinc on the performance and immunocompetence of broilers under heat stress. Poult. Sci. 82:1580–1588.

Boskey, A. L., T. M. Wright, and R. D. Blank. (1999). Collagen and bone strength. J. Bone Miner. Res. 14:330–335.

Crenshaw, T. D., E. R. Peo, A. J. Lewis, and B. D. Moser. (1981). Bone strength as a trait for assessing mineralization in swine: A critical review of techniques involved. J. Anim. Sci. 53:827–835.

Klasing, K. C. (1984). Effects of inflammatory agents and interleukin-1 on iron and zinc metabolism. Am. J. Physiol. 247:R901–R904.

Liu, D., H. P. Veit, J. H. Wilson, D. M. Denbow. (2003). Long-term supplementation of various dietary lipids alters bone mineral content, mechanical properties and histological characteristics of Japanese quail. Poult. Sci. 82: 831-839.

Sahin, K., and O. Kucuk. (2003b). Zinc supplementation alleviates heat stress in laying Japanese quail. J. Nutr. 33:2808–2811.

Sakaguchi, K, I. Morita and S. Murota. (1994). Eicosapentaenoic acid inhibits bone loss due to ovariectomy in rats. Prostaglandins, Leukoterienes, Essential Fatty Acids. 50: 81-84.

Sandoval, M., P. R. Henry, X. G. Luo, R. C. Littell, R. D. Miles, and C. B. Ammerman. (1998). Performance and tissue zinc and metallothionein accumulation in chicks fed a high dietary level of zinc. Poult. Sci. 77:1354–1363.

Schreiweis, M. A., J. I. Orban, M. C. Ledur, D. E. Moody, and P. Y. Hester. (2005). Validation of dual-energy X-ray absorptiometry in live White Leghorns. Poult. Sci. 84:91–99.

Shane, S. M. (1988). Factors influencing health and performance of poultry in hot climates. Crit. Rev. Poult. Biol. 1:247–267.

Velleman, S. G. (2000). The role of the extracellular matrix in skeletal development. Poult. Sci. 79:985–988.

Watkins, B. A, C. L. Shen , J. P. McMurtry, H. Xu, S. D. Bain. (1997a). Dietary lipids modulate bone prostaglandin E2 production, insulin-like growth factor-I concentration, and bone formation rate in chicks. J. Nutr.127:1084–1097.

Xu, H., B. A. Watkins, M. F. Seifert. (1995). Vitamin E stimulates trabecular bone formation and alters epiphyseal cartilage morphometry. Calcif Tissue Int. 57: 293-300.

Investigation of Egg Enrichment Using Organic and inorganic Supplements of Selenium and Zinc in aged Laying Hens

Reza Vakili, Javad Saebifar, Reza Majidzadeh Heravi

Abstract

This experiment was performed to investigate the effects of organic and inorganic sources of minerals zinc and selenium on yield, egg quality and selenium and zinc concentration in eggs in older laying hens. In this experiment, 210 Leghorn laying hens-line strains (w-36) at the age of 83 weeks were used based on a completely randomized design. Treatments included seven treatments with three replications and 10 chickens in each replication. Treatments included two inorganic sources of zinc and selenium including zinc oxide and sodium selenite and two organic sources of zinc methionine and selenomethionine and control treatment. Production yield, egg quality and selenium concentration on eggs were evaluated for six weeks. The results showed that organic and inorganic sources of zinc and selenium in the diet of laying hens did not have a significant effect on production performance and internal quality of eggs. Addition of zinc and selenium supplements from organic sources improved egg shell quality compared to the control group. Organic and inorganic sources of selenium and zinc increased selenium and zinc stores in eggs at 85 and 87 weeks ($P < 0.05$) .There was no significant difference between organic and inorganic sources for egg enrichment with selenium and zinc. Increased selenium and zinc storage was observed up to 88 weeks old laying hens. Addition of zinc and selenium supplements from organic sources improved egg shell quality.

Keywords: Egg, Selenium, Zinc, Egg Shell quality, Organic and inorganic Sources

The effect of zinc-containing diets on performance, egg quality and internal organs of laying hens

Reza Vakili Fathmeh Ashrafi

Abstract

The objective of this experiment was to investigation the effects of zinc-containing molting diets as alternative to the conventional feed withdrawal method, on performance,egg quality and the status of the internal organs of laying hens at 80 weeks of age. Experiment was carried out as a complete randomized design with 5 treatments, four replication and five samples in each replication. Treatments were include: molting diet+ 1% zinc oxide, molting diet+ 1.5 % zinc, molting diet+ 1% zinc acetate, molting diet+ 1.5 % zinc acetate, molting diet. Treatments were applied for 10 days to induce molting. Hens were weighted before and after molting, feed intake, egg production rate, egg production,egg quality, reproductive and non-eproductive organs weight and zinc accumulation in kidney and liver were measured. Data analyzed with ANOVA and the means were compared with Duncan multiple range test. Body weight loss and decline in feed intake were observed in all treatments. Egg production stopped after three weeks, in all treatments. After molting, egg production of birds on 1% ZnO egg production sooner than the other treatments. After molting, Egg shell breaking strength increased applying 1.5% zinc acetate. Ovary weight (as % of body weight) decreased using 1.5 % zinc acetate. Zinc bioaccumulation increased in liver and kidney applying zinc treatments. Force Molting with a diet containing 1% zinc oxide is recommended. Zinc-containing diets resulted to complete egg molting and improved egg shell quality. 1% zinc oxide resulted to faster return to egg production and less zinc bioaccumulation in the body.

Keywords: Force Molting , Feed intake , Egg quality ,Ovary weight, Zinc Bioaccumulation

ACKNOWLEDGEMENT:

This research has been financially supported by myself .

Publisher: Eliva Press Global Ltd

Email: info@elivapress.com

Made in United States
Orlando, FL
04 November 2026

90209050R00031